CHIMN
GABLES AND
GARGOYLES
A Guide to Britain's Rooftops

Trevor Yorke

COUNTRYSIDE BOOKS
NEWBURY BERKSHIRE

COUNTRYSIDE BOOKS
3 Catherine Road
Newbury, Berkshire

To view our complete range of books,
please visit us at
www.countrysidebooks.co.uk

ISBN 978 1 84674 354 2

Illustrations and modern photographs by Trevor Yorke

Produced through The Letterworks Ltd., Reading
Typeset by KT Designs, St Helens
Printed by The Holywell Press, Oxford

CONTENTS

INTRODUCTION

While we celebrate Britain's grand and spectacular architectural masterpieces there are many modest historic buildings in every town and city which receive less attention. Around famous city squares, along the high streets of our towns and down rustic village lanes are beautiful and interesting facades with decorative and fascinating rooftops, which are often ignored by passers-by. Part of the reason for this is that our attention is usually directed towards the ground floor from close quarters by glittering shop fronts and colourful signs. Also, many urban areas have been developed and trees have grown since they were built so the upper parts which were originally intended to be viewed as part of the whole facade are now obscured. It can therefore be an enlightening and surprising experience to look up and discover the architectural delights and historic structures which line the top of the walls and roofs of Britain's buildings.

This book is a celebration of these varied rooftop features which can be found on everyday buildings in our cities, towns and villages. More than that it explains what they are, why they were fitted, their changing styles and how they can help date a building. The book briefly explains how roofs have developed over the centuries before exploring the features you can see upon them. This includes towers and spires, parapets and balustrades, dormer windows and skylights, gable ends and pediments and not forgetting the spectacular chimneys which can be found on the rooftops of most period properties. It even features details like weathervanes, lead guttering, clocks, datestones and carved beasts glaring down from the ridge. Although all periods are covered there is an emphasis upon the Victorian when architects incorporated the roof into their highly decorative and colourful designs. The formerly plain gable ends suddenly became encrusted with complex timber patterns, polychromatic brickwork and beautiful patterned tiles while the very covering itself was formed into startling patterns edged by intricate ironwork, decorative bargeboards and ornate terracotta features.

Britain's rooftops are fascinating and often forgotten places waiting to be discovered. They can reveal the true historic value of buildings masked by modern facades and in many cases tell something about the ambitions of the people who erected them. Old signs, carved text and decorative motifs might show what a building was originally used for while the style of roof, form of decoration and design of chimneys can help date them. There will also be oddities which could be unique to a building and old features which have been retained while the walls below have been regularly updated and changed. So next time you walk down your high street or visit an historic place

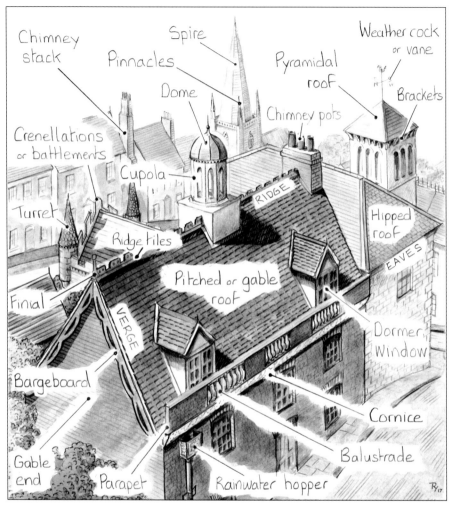

Chimney stack

Spire

Pinnacles

Dome

Pyramidal roof

Weather cock or vane

Brackets

Chimney pots

Crenellations or battlements

Cupola

Turret

Ridge tiles

RIDGE

Hipped roof

EAVES

Pitched or gable roof

Finial

VERGE

Bargeboard

Dormer Window

Cornice

Balustrade

Gable end

Parapet

Rainwater hopper

A view *over an imaginary town showing the variety of rooftop styles and decorative details which can be found, with labels highlighting the key features.*

take time to look up and study the details which could turn a seemingly ordinary building into something to be treasured.

Trevor Yorke

www.trevoryorke.co.uk

Follow me on Facebook at trevoryorke-author

5

THE IMPORTANCE OF BEING A ROOF
Styles and Materials

The roof is arguably the most important part of a building. Without it rain, snow and wind would penetrate inside making it uninhabitable and quickly ruining the structure beneath. At the same time it also keeps the heat in during winter and protects the interior from the effects of the sun. In traditional construction 'a good hat and strong boots', the roof and foundations, were regarded as the essential elements to ensure that a building would stand the test of time. As long as these were sound the walls beneath could simply be made of mud and straw and would last decades if not centuries. In addition to this protective role the roof is a prominent visual feature which can be used by the architect for dramatic effect. Its chosen form, the angle at which it is set and the colours and shapes of the covering material are a distinctive part of the architectural style of the building. Further interest is created by the decorative trimmings, dormer windows, towers, parapets, gables, gargoyles and chimneys which help create the lively skylines of Britain's towns and cities.

How a roof works

In our wet climate it has long been appreciated that a roof is best formed with a slope, the angle or pitch at

FIG 1.1: Looking up at buildings *can reveal architectural delights especially upon those from the 19th century. The 1891 Victoria Law Courts, Birmingham, pictured here, show how architects in this period used decorative towers, finials, chimneys, gables, cornice and dormers to create a lively roof line which can easily be seen from below.*

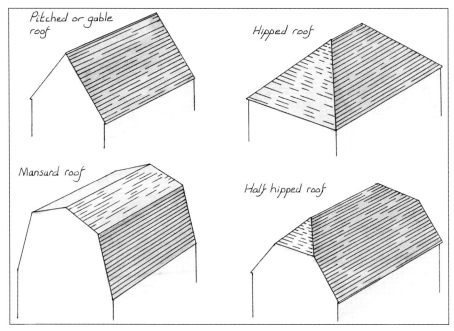

FIG 1.2: The most common *forms of British roofs. Large buildings can be covered by a complex arrangement of slopes but still tend to be composed of these primary forms.*

which it is set being determined by the covering's efficiency in shedding rainwater. A porous material like thatch had to be set at a steep pitch so that the rain ran off or evaporated before it could soak through the layers of straw or reed. Large Welsh slates on the other hand had few gaps for water to get through and could be set at a shallow pitch. The effects of the wind also had to be accounted for as it would push down on the windward side and suck up the covering on the leeward. The pitch of the roof and the way in which the tiles or slates were fixed ensured a gale would not cause damage. In addition there was the dead weight of the roof structure and covering to consider, which could be greater in winter when snow was laying on top. A heavy covering like plain clay tiles or stone slates would require thicker or more tightly packed supporting timbers irrespective of the pitch of the roof.

There was another problem which affected pitched roofs. Just as when you rest two playing cards at an angle against each other on a flat surface they are prone to slip and fall flat, so the sloping sides of a roof are always trying to spread out. The flatter the angle and heavier the roof the more they apply a horizontal thrust and try to push the tops of the walls outwards. Masons, carpenters and architects had to allow for this effect by adding exterior

7

FIG 1.3: Ely Cathedral, Cambs: *Buttresses help counter the horizontal thrust from the pitched roof so the walls can be thinner and filled with glass. The flying buttresses pictured here are used in larger churches to span the side aisle with the decorative pinnacles on top acting as weights to add stability to the supports.*

FIG 1.4: The straw or reeds *are laid from the eaves upwards with the bundles spread out over the battens and temporarily held by reeding pins (1). The surface is then dressed with a legget, a wooden bat with a ridged surface, to form a smooth even surface (2). This is held by sways fixed into the rafters (3). A ridge piece is formed along the top and the eaves trimmed to complete the roof.*

FIG 1.5: A limestone slate roof *from the Cotswolds. The stepped slates projecting out are a traditional method of keeping rain away from the junction between the roof and the wall. Note the slates are larger along the eaves (bottom) and smaller at the ridge (top).*

supports like buttresses or form trusses inside which tied the sloping sides of the roof together. This has placed restrictions upon the size and plan of buildings and their design has been in part shaped by the need to counter the effect of this spread.

The style of roof

The form of the roof that you see when looking up at buildings has been shaped by changes in architectural fashion and the types of roof coverings used. The most common covering on medieval buildings was thatch (from the Old English *thaec* meaning roof covering), although straw, reeds and other vegetation fell from favour in many towns from an early date as it allowed fire to spread easily. In rural areas its use continued into the 19th century but by this time it was associated with poverty, covering the cottages and hovels of the poor. Today there are three types of materials used for thatching. Water reed has been traditionally limited to

FIG 1.6: Large medieval, *Tudor and early Georgian buildings were usually spanned by a series or pair of short spanned roofs as here at the Guild Hall, Thaxted, Essex.*

the Norfolk Broads and a few coastal areas where they grew in sufficient quantity although much of what is used today is imported. The ends or butts of the reeds are flush so the finished surface appears smooth but the ridge section is usually formed out of straw as it is more flexible. Combed wheat reed is straw which has been specially prepared for thatching by removing the waste so only the stem remains. Long straw was the most common type in the past with the material being used uncombed with the heads still attached. It is laid out and wetted to make it more pliable and is then gathered into yealms, the bundles which are carried onto the roof to be laid. Long straw has lines of liggers (thin wooden rods which hold down the thatch) around the eaves and verges and not just the ridge as with reed. As just the top layer is replaced (coatwork) rather than a complete new

FIG 1.7: Traditional cast lead *(top) can last centuries, its durability coming from the salts in its composition which are insoluble and form a protective coating as it weathers. This example from St Mary's church, Melton Mowbray records its restoration in 1867. Copper (bottom) was often used on towers and domes. It gains a distinctive bright green coating once it is exposed to the atmosphere when acidic pollutants create a chemical reaction which forms copper sulphate. It is this green patina which prevents further weathering.*

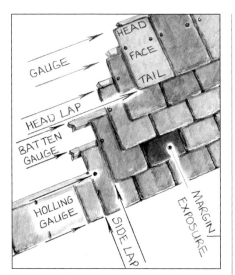

FIG 1.8: Cut away views *of a double lap slate roof (top) and a single lap pantile roof (bottom).*

FIG 1.9: The roofs of some *late 17th and early 18th century buildings featured deep sculptured cornices, white painted dormers with alternate triangular and arched pediments, and a central cupola or lantern as in this example from Bury St Edmunds.*

covering there are some examples of medieval thatch still in existence under later thatch, with their undersides still blackened by smoke from open fires over 500 years ago.

Some parts of the country were fortunate to have sedimentary stones which could be split into thin slabs (fissile) which are known as either tiles, tilestones or slates. These were used from an early date on important buildings but did not become common until the 17th and 18th centuries. In parts of Yorkshire and Derbyshire sandstone was used, which could be naturally split or sawn in a quarry to make stacks of thin slates. These were worked into shape by the knapper or striker who would hammer the edges over a vertical stone and then sort the finished pieces

11

FIG 1.10: Traditional plain *clay tiles were hand-made and have a bowed profile. Later machine-made types are more regular and square edged. Although the tiles appear like small rectangles only a small portion is exposed as they are double lapped (see FIG 1.8).*

into set sizes which were known by distinctive local names. Limestone roof tiles or slates are a distinctive feature of the Cotswolds and Northamptonshire. Those which naturally formed slabs of even thickness ready to be trimmed to size were referred to as presents, those which had to be split were called pendles. Presents were used as far back as Roman times but it was not until the 16th century that an effective method for splitting limestone to form pendles was discovered. On most stone roofs the tiler would have a random selection of sizes available and would fit the largest ones at the bottom of the roof over the wall and graduate up to the smallest at the ridge. This was because the larger tiles were thicker and hence heavier so

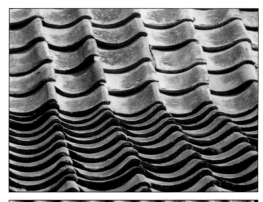

FIG 1.11: Traditional clay *pantiles from Norfolk (top right) with their distinctive flat 's' shape profile which covers up the gap with the adjoining tiles so they can be laid single lapped (see FIG 1.8). Glazed black pantiles (bottom right) were an alternative which were used in the 18th and 19th centuries.*

FIG 1.12: It was common *in the first half of the 18th century for large houses to be spanned by a pair of parallel steep pitched roofs as in this example. Note the parapet along the top of this Classical style building hides the roof from the street.*

were placed directly above the walls for the best support.

As these early coverings were porous and had lots of gaps to let water through they were set at a steep pitch. Hence most medieval and Tudor buildings were composed of a series of narrow structures which suited these forms of roof covering and avoided the need to build complex timber trusses inside. Even large buildings were often formed from a line of single depth rooms around a central courtyard so as not to stretch roof technology too far. When a wider space like a church was spanned, oak trusses or external buttresses needed to be added. Lead, which provided a seamless, watertight surface and could be set at a shallow angle, was also used, especially in the 15th and 16th centuries, despite its expense.

In the wake of the Great Fire of London in 1666, regulations concerning the roof covering came into force in the capital and later elsewhere. Combustible materials like thatch and shingles were banned and stone slates and clay tiles became common. This latter form had been used from the 13th century in certain eastern and southern areas and were usually made on site from local clay. Production became more widespread from the late 17th century as permanent brick and tile works were established to meet the new urban demand. Although only a small rectangular section is visible due to them being double lapped, each plain tile is approximately 10½ x 6½ inches (265mm x 165mm), a standard size established in the 15th century. To help prevent capillary action allowing rain

13

FIG 1.13: Regency buildings *are characterised by low-pitched hipped roofs as large Welsh slates became widely available.*

or snow to work up between tightly fitted tiles they were usually made with a slight camber giving a gentle rippled effect to the finished surface. They traditionally had two holes near the top through which oak pegs were pushed to hook over the horizontal battens, later machine-made tiles usually had nibs formed at the top edge instead.

In the late medieval and Tudor period Britain was exporting large quantities of wool and cloth from the eastern counties to the Low Countries and in order to ballast the returning ships they were loaded with roofing tiles which had a distinctive 's' shape profile. This form meant that one side could hook over the adjacent tile and

FIG 1.14: Grey slates *with a blue, purple and green tinge capped by terracotta ridge tiles are a distinctive feature of Victorian towns and cities as in this view in Chester.*

cover the gap between them so they could be more widely spaced out and set at lower pitches. These pantiles as they became known were a distinctive feature of roofs along the coastal region from Scotland down to the Thames and also around the Severn Estuary. Early examples were used on less important buildings like outhouses or barns, but as domestic production began and standards improved from the late 17th century they came into more general use.

As Classical styles became fashionable during this period so architects sought to emulate the buildings from Ancient Greece and Rome and the steep pitched roof rather spoiled the effect. Hence parapets were fitted around the top of

FIG 1.15: In the 1870s *the French Empire style was popular, with square plan domes (as above), steep pyramidal roofs and richly decorated round dormer windows distinctive features of this style.*

FIG 1.16: The roof *with its decorative details was a key part of the design of many Victorian buildings. This superintendent's house at Papplewick Pumping Station, Notts is enhanced by the staggered skyline and low-slung roof forming a porch over the door, corresponding to that over the bay window on the right, so that structure presents a different form on each front.*

15

FIG 1.17: Many late *Victorian and Edwardian buildings featured prominent chimneys and timber-framed gables as in these examples from Leek, Staffs.*

FIG 1.18: Single piece *clay tiles with one or two ridges and interlocking edges were mass produced in the early 20th century and marketed as Roman tiles (top) as they imitated the appearance of those used across the Ancient empire. Bright green glazed pantiles are a distinctive feature of modern style properties from the 1930s.*

the walls to hide the offending tiles or slates from view from the street below. From the late 18th century, true slates from Wales and other parts of the country began flooding the market and helped to transform roofing in the Regency and Victorian periods. It is a metamorphic rock which readily splits into thin sheets, created from ancient clay and mud which was transformed into shale and then, due to heat and pressure, was chemically altered to form slate. The stone was split (cleaved) in the quarry and then dressed by hand to form the different sizes and shapes, with holes punched through so they could be nailed to the battens. As a result of this new material and a better understanding of forming trusses, large buildings could be covered by a single roof with a shallow pitch. During the first half of the 19th century hipped roofs, with all four sides sloping, became fashionable especially on Italianate style buildings which had a deep overhang supported on richly decorated brackets. As the number of household servants increased during the 19th century and bedroom accommodation for them was provided in the loft so the mansard roof which created more headroom inside became a common feature on large terraces in towns and cities.

From the 1840s, Gothic Revivalists changed the way in which roofs were designed. Parapets began falling from favour and architects made the roof a key feature of their designs. The pitch became steeper once again with some covered in patterned or coloured tiles and capped off by fancy ridge pieces. Dormer windows and gable ends with decorative bargeboards were prominent parts of the design while the roof line was broken by spires, towers and prominent chimneys. In the 1870s and 1880s, flamboyant French Empire style roofs were popular with their straight or curving slopes often covered in diamond or fish scale shapes and round dormers providing light inside. A revival of late 17th century houses made Dutch gables and little white painted dormers a key part of this new Queen Anne style. In the last decades of the century, architects working under the banner of the Arts and Crafts movement created long sloping roofs with eyebrow or low dormer windows.

In the 1920s and 1930s, steeply pitched hipped roofs covered in clay tiles were a common feature in the rapidly expanding suburbs although modernists occasionally interrupted these with bold geometric, white houses with flat roofs where the optimistic could take in the sun's rays. In the wake of the Second World War, chronic shortages of materials forced builders to simplify designs and expensive hipped roofs

FIG 1.19: Corrugated iron *(from the Latin word 'ruga' meaning wrinkled) is perhaps one of the most common roofing materials around the world. It was patented in 1829 by Henry Robinson Palmer, resident engineer at London Docks, however he was more interested in his pioneering monorail system and sold his invention to one of his contractors, Richard Walker. Corrugated iron was used for railway station roofs, agricultural buildings and to replace thatch on cottages. Prince Albert even ordered a corrugated iron ballroom for Balmoral, which still stands today. British manufacturers also exported complete buildings around the world from cottages to churches like this example at Avoncroft Museum of Historic Buildings, Worcs.*

became a rarity. Lower pitched roofs increasingly covered with concrete tiles became common in the housing market by the 1960s while larger buildings used concrete, glass and metal sheeting.

17

THE DECORATIVE TOUCH
Dragons, Ridges and Ironwork

The structure and covering of the roof provided the buildings of Britain with weatherproof protection. Although many regarded it as a functional element there have been times when the roof has been decorated, as craftsmen and architects have recognised its aesthetic value and displayed their art by creating patterns or figures upon its sloping sides and ridges. This is particularly true of the Victorian period when the Gothic Revival inspired architects to use decorative patterns and colour to enhance the functional parts of a building rather than just applying ornament in a haphazard way. Hence from the mid 19th century through to the outbreak of the First World War roofs became emblazoned in all manner of patterns, terracotta ridge tiles and decorative ironwork.

FIG 2.1: The glorious and eclectic pumping station at Abbey Mills, East London is capped by decorative iron finials and crests; a popular finishing touch from the 1860s and 1870s. Improvements in the production of cast iron during the late 18th and early 19th centuries enabled Victorian iron foundries and builder's merchants to publish catalogues full of designs including decorative finials for the end of the ridge, weathervanes and fancy short railings to line the ridge or crest of towers.

Thatched ridges

The crest along the top of the roof is the most vulnerable part of the covering and

FIG 2.2: The ridge *needs to be replaced more often than the main body of the thatched roof because the wooden spars and liggers (top) rot. Although a well formed ridge can last for decades many owners prefer to replace them every 10 to15 years to keep the roof looking tidy and secure (bottom).*

has always required special treatment to ensure it stays watertight. On thatched roofs the ridge is usually formed from straw or grass (water reed is too brittle) around a long pipe of thatch called a dolly. The ridge is pinned in place by long wooden liggers which are fixed by U-shaped spars forced into the thatch. These liggers and spars are arranged in patterns which add a decorative touch to a thatched roof. Old photos and paintings show cottages with the ridge finished flush with the main roof but a more decorative method which has become widespread over the past century are block ridges where an extra layer of straw is pinned down and then cut into a straight line or a pattern of scallops and diamonds to make a decorative raised edge to the ridge.

FIG 2.3: A drawing *of a wrapover ridge being formed showing the dollies, liggers and spars.*

FIG 2.4: Animals and birds *often feature as a finishing touch on some thatched roofs. Chicken wire is usually fitted over the roof to stop birds pulling the straw out and to help hold down the spars and liggers although it is often not fitted on reed as these tend not to be an issue.*

Decorative tiling and ridge pieces

Gothic Revival architects inspired by the writing of Augustus Pugin and John Ruskin began designing the structural parts of their buildings with decorative designs formed within the brick and stone. From the 1850s to the 1870s and sometimes later the walls of buildings could become a riot of colour and geometric patterns. This fashion extended to the roof which was a prominent feature on Gothic

FIG 2.5: Examples of *Victorian tiled and slate roofs with decorative bands formed from scalloped, fish scale and diamond shapes. Patterns could also be formed from different colour tiles as in the glorious example bottom centre.*

FIG 2.6: The vulnerable ridges *and sloping hips of roofs were usually covered with special shaped clay tiles. Angled (top) or half round (centre) were common and could be continued down the corners of a hipped roof. In the 1920s and 1930s bonnet tiles were often used on the latter feature (bottom). To prevent the tiles slipping, a vertical strip of iron with a curly tip known as a hip iron, was fitted at the bottom (centre).*

FIG 2.7: Examples of *Victorian and Edwardian decorative ridge tiles.*

21

Revival buildings. The most common arrangement was to form horizontal bands of scalloped or profiled clay tiles or diamond slates within the plain surface. Alternatively a roof could be broken up by rows of tiles or slates of contrasting colour. The ridge was another focal point for decoration by Victorian builders eager to raise the status of their work. Clay or terracotta moulded ridge tiles were available with a wide range of patterned crests. Those on mid-19th century buildings tend to be the fancier; by the turn of the 20th century they are generally plainer. Short decorative ironwork railings were popular especially in the 1860s and 1870s to line the crest of roofs and trim the top of towers.

Finials

A finial is a decorative vertical feature, in the case of a roof standing at the terminal of the apex or ridge (the word is derived from the Latin '*finis*' meaning end). They are believed to have been fitted to Roman roofs and were certainly a feature of some in the medieval period. It was the Victorians who took the finial to a new level by featuring them above the gable ends of buildings large and small. There were simple wooden pinnacles, intricate ironwork models and occasionally carved stone figures. The most common material though was clay or terracotta which could be moulded into finials featuring balls, spikes, fleur de lys, and foliage designs. Another popular form was mythical

FIG 2.8: Simple stone ball *finials (top left) appeared on many 17th and early 18th century buildings. Later Victorian types ranged from simple terracotta pinnacles (top right) to flowing foliage designs (bottom left), while religious buildings usually had a crucifix (top centre). Some grand buildings had elaborate creations like the lantern type (bottom right).*

FIG 2.9: Mythical beasts were a popular feature at the end of the ridge. Dragons were a particular favourite of the Victorians and are still produced in a wide variety of designs today.

FIG 2.10: Railings along the crest of roofs and towers (left) were popular in the 1860s and 1870s but the elaborate iron finials have a wider date range.

23

creatures like gargoyles and dragons inspired by medieval designs which traditionally were believed to ward off evil spirits and ensure good luck.

Weathervanes and weathercocks

Weathervanes, from the Old English word '*fane*' meaning flag, are a distinctive feature of rooftops and are recorded as far back as the ancient world. The Vikings fitted quadrant shaped types on the front of their lead ships as a navigation aid and some of these were later fitted to church towers in Scandinavia as the nation's ocean travel declined in the 11th century. The familiar weathercocks with the vane shaped as a cockerel are supposed to have

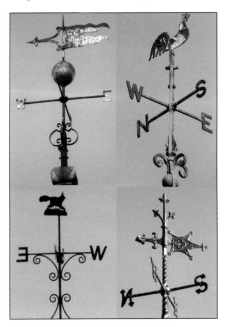

FIG 2.11: Examples *of weathervanes or cocks which, like iron finials, could also act as lightning conductors (top left).*

had their origins in a Papal edict from the 9th century. The Pope decreed that every place of worship should display the bird to remind the congregation of Peter's betrayal of Christ in the Gospel of Luke when he warns him: '*I tell thee, Peter, the cock shall not crow this day, before that thou shalt thrice deny that thou knowest me*'. There are records of a weathercock at Winchester Cathedral in the 10th century and one is shown being fitted onto Westminster Abbey in the Bayeux Tapestry. During the late medieval period a wider range of animals were used including fish, a traditional Christian motif; griffins, a symbol of strength and vigilance; and the dragon, a warning against sin. The wealthy also fitted vanes on top of their castles and private residences in the form of a trailing banner displaying their coats of arms, a fashion probably descended from the fabric banners which archers used in battle to find the direction the wind was blowing. Separate pointers indicating the direction of the wind did not become common until the 17th century by which time wrought iron versions with a date stamp or a flat silhouette figure were increasingly popular. The Victorians mass-produced cast iron types with all manner of decorative symbols and motifs and these are still a common sight on a wide range of buildings.

Lightning rods

Lightning strikes on tall buildings like church spires have long been an issue of concern. In 1749 when Peter Ahlwardt, a German theologian and philosopher, issued advice on dealing with thunder

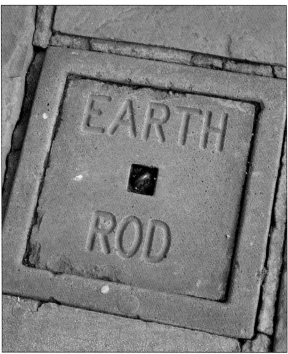

FIG 2.12: Lightning rods *connected to the ground by a strip of wire protect tall structures from strikes (left). The earth rod often has a cover in the ground to provide access (right).*

and lightning he simply recommended that during a storm everyone should keep as far away from a church as possible. However, shortly afterwards the American scientist and politician Benjamin Franklin invented the lightning conductor during his experiments with electricity. He suggested that an iron rod with a sharp point should be mounted at the highest point on a building and connected via a wire or strip of metal to another rod sunk in the ground which would earth the lightning strike. His British compatriots thought that the pointed end would attract more lightning so recommended

the rod should have a blunt end; the use of the former in the colonies became a political statement in the run up to the American War of Independence. One problem with the conductor was that it was not clear when it had been struck so damage to the building or the conductor could be missed. To prevent this hollow glass balls were fitted on the rooftop rod which would smash if it was hit and warn residents that there had been a strike. During the 19th century lightning rods were often decorated or incorporated with a weathervane, both featuring colourful glass balls on their shafts.

25

FIG 2.13: Due to the risk of fire *in many industrial buildings experiments were made with cast iron roofs. In 1827 Elias Carter from Devon patented a roofing system using cast iron plates and a complete roof still survives at The Crescent in Gloucester Hospital which was installed after a fire in 1832. A far more famous building also has a cast iron roof. The original Palace of Westminster was destroyed by flames in the 1830s so fire prevention was high on the list of requirements for its replacement designed by Charles Barry and Augustus Pugin. The new roof on the Houses of Parliament was built with cast and wrought iron trusses supporting metal purlins and rafters upon which square pan-shaped cast iron tiles were fixed (top left). These were up to 39 inches (1 m) in length, ½ inch (1 cm) thick and could weigh up to 165 lbs (75kg) each. After 160 years and despite some leaking the main structure is still sound and recent restoration has further extended its life. At the restored boiler house of Papplewick Pumping Station, Notts dating from 1881, slates were tied by wire to a framework of iron angled battens, 'T' shape rafters and beams with the underside torched to make it more fire retardant. Unusually the decorative ridge and hip tiles on the exterior surface were made from cast iron (top right).*

BRING ME SUNSHINE

Dormers, Lanterns and Vents

Most medieval and Tudor buildings were a single room deep, partly to keep the roof structure simple. This limited depth meant rooms could have light entering both sides and cross passages allowed fresh air to ventilate the interior. However, as buildings generally became deeper from the 17th century, parts of the interior had limited light and became intolerably dark so glass lanterns or skylights in the roof began to be fitted. Also as the demand for interior space increased so the formerly empty loft was converted into accommodation and vertical panes of glass capped by a miniature roof, called dormer windows, were fitted to illuminate it. In other situations there may have been a need to ventilate the interior and roof-mounted louvred openings could help draw air out from inside.

FIG 3.1: Dormer windows, *skylights and towers with vents are a key feature of the roofs of large buildings. As structures grew in size, sunlight and fresh air became short in supply so architects created openings on top to allow it inside. This became an important part of the design of industrial and public buildings from the 19th century and the Victorians treated these additions as another feature which they could decorate.*

Roof vents

Medieval buildings from large halls down to small houses used open fires to provide heating and cooking. Before fireplaces and chimneys became popular during the 16th and 17th centuries the smoke produced would either drift up to the roof timbers and out through the thatch or exit via a louvred vent opening in the ridge of the building. Roof vents continued to be important in later industrial buildings where boilers and machinery produced excessive heat and rooftop louvred vents were often fitted. In the 19th century a wider range of mass produced versions were produced and where they survive today it can give a clue to the original use of the building.

FIG 3.2: **In buildings** *containing groups of people, like hospitals or schools, large vents were often fitted along the roof to circulate air. At the former Spring Hill School, Birmingham, pictured here, the spire and tower serve as a vent for a system designed by the architect Joseph Chamberlain to draw fresh air through the buildings.*

FIG 3.3: **A beautifully** *carved vent on the roof of the former public library at Newark, Notts, which was opened in 1883.*

FIG 3.4: Medieval buildings *often had simple vents or louvred openings along the ridge (top left) before chimneys became standard. Cupolas, vents in gable ends and short towers with spires featuring louvred openings were common on large public and commercial buildings in the 18th and 19th centuries to help circulate air and remove heat. The presence of a fitting such as those illustrated above can help identify the former use of a building.*

Skylights and lanterns

A lack of light became an issue in many buildings as they grew in size. In the 12th and 13th centuries growing congregations forced medieval churches to be expanded by adding aisles down the sides, which made the central nave darker. One solution was to fit a clerestory, a row of small windows just

FIG 3.5: Weak foundations *were assumed to have caused the collapse in 1322 of Ely Cathedral's original tower and as a result its replacement was wider so as to avoid the soft ground. To keep the weight down the new structure was capped by a timber and lead lantern which let colourful sunlight stream down into the formerly dark centre of the cathedral. Another way of getting light into these wide edifices was to install a row of windows along the top of the nave called a clerestory. This often happened when a new flat lead roof was installed and you can just see on the right of this photo the marks left on the tower where the former steep pitched roof originally stood.*

FIG 3.6: Different types *of skylights and clerestories on Victorian commercial buildings to help illuminate the workings inside. Note the maltings (bottom) also has louvred vents in the walls.*

FIG 3.7: The Gothic *skylight illuminating the central staircase at Strawberry Hill House, Twickenham. Although skylights became very fashionable from the late 18th century most were simple frames of glass and not as elaborate as this example in Horace Walpole's 'little castle'.*

under the eaves of the nave. Another was to make an opening in the roof and cover it with a tall glass and timber structure called a lantern. These can be found from the late 17th century often with decorative ironwork weathervanes fitted on top. Similar structures were also built as observation towers, especially in seaside towns or popular tourist destinations during the 19th century.

In some Georgian buildings a central staircase was a convenient part of the fashionable symmetrical plan so architects began installing skylights in the ceiling above to cast light down the steps. These were a luxury fitting which were designed for each specific building but during the Regency period improved glass and iron production made standardised types cheaper and widely available and they became a distinctive feature of large urban houses during the 19th century. Many of these also had sections which could be opened to provide the ventilation which was often required above a service room in a rear extension. Others had wonderful stained glass patterns creating a sparkling kaleidoscope of colour in the space below.

Dormer windows

The roof timbers in most medieval halls were open to the floor below where the owner and his household could eat, drink and sleep. As the owners of houses began to desire more privacy later in the period and they started engaging wage-earning servants so large houses began to be divided up with floors and ceilings. The loft space which was created would become an ideal space for staff sleeping quarters and small vertical windows capped by a miniature roof were fitted in the slope of the roof to provide the necessary light and ventilation. It was really from the second half of the 17th century that these dormer windows became a common feature at all levels of society. In large urban houses where space was at a premium the loft was an essential part of the plan and they were illuminated by white painted dormers with semi-circular and triangular pediments above. These became less ornate in the 18th century and were

FIG 3.9: Dormer windows *from late 17th to early 19th-century buildings. Early types had lead casement windows (top left) but sash windows became common during the 18th century with horizontal sliding types (top right) on some cottages.*

FIG 3.8: Dormer windows *were a common feature of 18th and 19th-century cottages where costs could be kept down if the building was just a single storey with the bedrooms built into the roof space.*

FIG 3.10: Early Victorian dormers usually had sash windows (top left) but casement windows became fashionable later (top right). Queen Anne style houses often had shaped gables (bottom left) while Arts and Crafts buildings had low, wide dormers with stained glass (bottom right).

partly obscured by parapets so as not to spoil the proportions of the fashionable Classical facade.

As the roof once again became an architectural feature in the mid-19th century so the dormer window provided an opportunity for decoration to complement the style of the building. Gothic style dormers tended to have steeply pitched little roofs above them with fancy carved bargeboards fitted along the gable. Queen Anne style buildings could have shaped gables comprising of quadrants, right angles and triangles. Edwardian types tended to be less ornate but can be found in a wide range of styles often with a revival of leaded glass or small paned windows. After the First World War, difficulties in recruiting servants and the fact that new houses were built on cheaper suburban land and hence were larger meant there was little need to use the loft and dormers fell from fashion.

33

FIG 3.11: **The revival** *of domestic styles of architecture championed by Arts and Crafts architects resulted in beautifully carved timber dormers (left) and imaginatively designed geometric forms, deep eaves and long, low windows (right). Note the vertical slits in the gables which are vents for the loft space and are distinctive of Edwardian properties.*

FIG 3.12: **In the** *16th and 17th centuries, most glass was blown and spun at the end of a tube to form discs, which could only produce small diamond panes of glass. The Georgians devised methods which enabled larger rectangular pieces to be made for windows and skylights but taxation and the cost of polishing made them expensive. It was not until the 1830s and 1840s that the industrialised production of glass and removal of taxation revolutionised its use. Mass-produced cylinder glass, in which a long tube of molten glass is cut along its length to form a flat sheet, and then plate glass, which was rolled on a cast iron bed, became available and helped make the material a cost-effective and lightweight roof covering. Paxton's Crystal Palace for the 1851 Great Exhibition with its distinctive iron and glass structure displayed the architectural possibilities of glass and over the following decades it was widely used to roof railway stations and shopping arcades as pictured here. Since the 1960s float glass, in which the molten material is poured onto a bed of liquid tin creating a perfectly smooth finish, has dominated the market.*

A DRAMATIC SKYLINE
Towers, Turrets and Spires

Perhaps the most striking features of any town or cityscape are the towers, spires and domes which break the horizontal lines of roofs. The great medieval cathedrals, Classical public buildings and Victorian Gothic churches create a lively vista lacking from many modern urban areas. Where would Oxford be without its dreaming spires or London if the dome of St Paul's had never been built? These great historic buildings inspired the designers of the more modest structures that line the streets of British towns and cities to fit more compact versions. However changing architectural styles and fluctuating religious endeavour have resulted in towers, spires and domes falling in and out of fashion.

Castle towers which combined a military role along with a statement of power, and church spires pointing towards heaven as a sign of faith were key features of these medieval buildings which continued to inspire masons and architects in later centuries. Many Tudor and Jacobean buildings featured towers but they were capped

FIG 4.1: Spires, *towers and domes break the skyline of this view across the Victorian mill town of Leek in the Staffordshire moorlands.*

with distinctive onion shaped roofs. A dramatic skyline was a key feature of the Baroque style which became fashionable in the late 17th and early 18th centuries so towers and domes adorned the finest country houses and public buildings. A cupola, a small dome set on a circular or polygonal base, was a popular rooftop feature during this period and variations of them remained common on public buildings, factories and stables into the early Victorian period. Spires were used by the eminent architects Sir Christopher Wren and James Gibbs for their London churches.

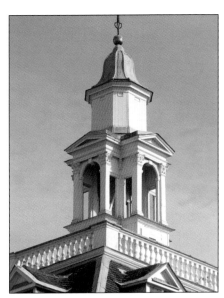

FIG 4.2: The stone keep *at the heart of the Tower of London was built for William the Conqueror but the distinctive onion shaped domes on the corner towers were a Tudor addition.*

FIG 4.3: Cupolas *were a distinctive rooftop feature of many public and commercial buildings in the late 17th century, often in conjunction with white painted railings and dormers (top). They can also be found throughout the 18th and 19th centuries housing bells and clocks.*

These were made in an imaginative variety of forms, most sitting on top of an open lantern, belfry and clock tower and together forming a steeple which became a popular feature on churches and public buildings in the 18th century on these shores but more notably in the United States of America.

The Georgians became obsessed with the Ancient world and their buildings were inspired by Roman and Greek temples and public buildings, with their architectural focus upon the Classical orders, symmetry and proportion. Porticos and pediments were the key points along their horizontal facades and towers and spires fell from favour with only domes occasionally breaking the horizon. During the more architecturally diverse period of the Regency some buildings were adorned with domes and towers, inspired by domestic medieval forms and others from Europe and Asia. During the 1840s Gothic Revivalists inspired a passion for towers and spires at first as a feature of religious buildings but by the 1860s as a more general architectural addition.

FIG 4.5: Many Victorians *imagined the towers of castles were capped by tall conical roofs, mainly because the few studies made into these medieval buildings had been on the Continent where this form had been popular. When William Burges restored Castell Coch near Cardiff in the 1870s and 1880s he used such roofs for architectural effect rather than historic accuracy despite trying to convince his critics otherwise. He even fitted a circular tower with a conical roof on his private home in London, pictured here.*

FIG 4.4: James Gibbs' *steeple on St Martins in the Field, London, dating from the 1720s.*

FIG 4.6: The Victorians *loved towers and they were a key feature of large public and commercial buildings. Inspiration came from sources across Europe as in these examples from Congleton (top left), Manchester (top right), Lockerbie (bottom left) and March (bottom right).*

FIG 4.7: Smaller towers *were integrated into more modest buildings during the 19th and early 20th centuries. Classical (top left) and Elizabethan (top centre) styles were still popular in the 1840s. The Italianate with a low pyramidal roof and round arched openings (top right) were fashionable in the 1850s but towers based upon those on medieval churches and castles were common from the late 1850s to the 1880s (centre left). Towers were often incorporated into the inner (centre) or outer (centre right) corner of many Victorian buildings. In the Edwardian period Baroque style towers were popular (bottom left and centre) and by the 1920s plain Art Deco types with a stepped crown became a distinctive feature of inter-war buildings.*

39

This fashion filtered down to more modest structures and stubby towers and pointed turrets are a common feature of urban buildings throughout the Victorian period. The Edwardians continued to enliven the skyline of their towns and cities with towers fitted into the roof but they had a particular passion for domes inspired by the 17th century Baroque.

FIG 4.8: Clocks *began to feature on cupolas and towers from the late 17th century when the pendulum clock was introduced, usually with just a single pointer indicating the time. From the early 19th century cheaper brass and iron reduced the price of these turret clocks and most then had separate pointers for the minutes and hours. They were a common feature on stable blocks, barracks, churches and factories where timekeeping was important but declined in use from the First World War as pocket and later wrist watches became affordable. Time was originally set locally by observations from a sun dial. It only began to be established on a national basis from Greenwich, at first on the railways, from the late 1840s and then internationally from 1884.*

FIG 4.9: The top section *of some towers seems mainly to have been built as a private room with a view. This might have been part of an exclusive development, a seaside hotel or just on the whim of the owner. Some were a simple cupola type structure; the example (top left) had a balcony around the outer edge. Others were the upper stage of a corner tower with spires and candle snuffer type roofs popular in the second half of the 19th century and domes more common in the early 20th century.*

FIG 4.10: Turrets, *small cylindrical towers projecting from the wall, were a common feature anytime a castle was the inspiration for a building. They were particularly popular in the late 19th century when the Scottish Baronial style was in fashion and usually featured tall conical roofs (left and centre) and occasionally domes or ogee shaped caps (right).*

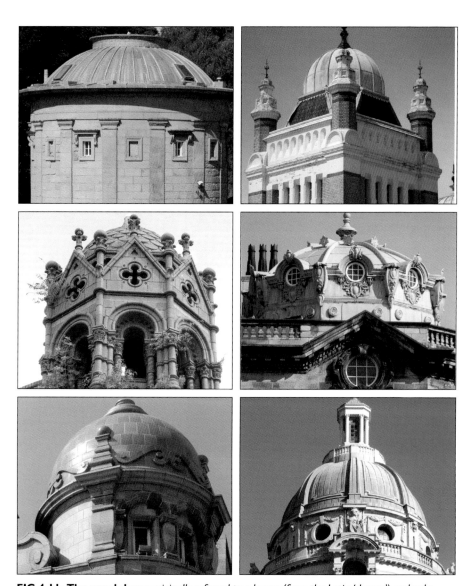

FIG 4.11: The word dome *originally referred to a house (from the Latin 'domus') and only began to be used to describe a half-sphere shaped structure from the 17th century. They were a distinctive feature of some of the finest Baroque buildings and shallow versions were popular in the Regency period (top left). The Victorians occasionally used them, especially when giving a building an exotic flavour (top right and middle left). The Edwardian revival of the flamboyant Baroque resulted in the widespread use of domes (middle right and bottom left and right).*

FIG 4.12: Spires *(from the Old English word for shoot or stalk) are a distinctive feature of medieval and Victorian churches. Medieval builders found an octagonal form was best for these structures but they fitted uncomfortably upon a square planned tower, hence from the late 13th century they developed the broach spire which had small angled broaches to cover the gap at the bottom of the diagonal sides (left). Later the spire became narrower and was set behind a parapet (centre left). The small openings down the sides are spirelets which ventilate the inner structure, earlier types tend to be larger than later ones. Constructing masonry spires was particularly tricky especially where the faces of the spire were flush with those of the tower. They were built from within around an internal timber scaffolding until near the top the masons would have to complete the structure and cap it off from the outside. Later types set behind a parapet were potentially easier to build as ladders and scaffolding could rest upon the top of the tower. Salisbury Cathedral, still the largest medieval spire in Britain, has a timber framework inside which is believed to have been inserted to reinforce the structure after a storm in 1362. A windlass, a large wheel with handles which used to haul up stones during its construction around 50 years earlier, is still in situ. The Victorians added spires to their new churches, at first remaining faithful to medieval types but later in the period experimenting with more ambitious designs from a wider range of sources (centre). Two popular types which can be found on smaller churches from this period are the splay footed spire (centre right) and the Helm roof (right) which had a gable on top of each wall of the tower and four diamond shaped sides to the spire-like top. The finest spires were built in masonry, other types were often covered in slates or shingles. Some were covered with lead which unfortunately for the builders of St Mary's and All Saints church, Chesterfield (centre left) has twisted and bent the spire in the 650 years since it was erected, due to the sun heating one side and inadequate timber bracing inside.*

THE TOP OF THE WALL

Parapets, Balustrades and Cornice

There have been times through history when, due to demands of defence, fire safety or architectural fashion, the roof has been hidden behind a parapet. The word originates from the Italian *parare* which means to defend or cover, highlighting the original military role of a raised wall around the edge of a roof. However even in the ancient world parapets were prescribed as a measure which could reduce the chance of flames setting fire to the roof and adjacent properties, and this was put into legislation in more recent centuries in many British towns and cities. Architects have also appreciated that raising the exterior wall obscures the roof from view and plain, panelled or fretwork parapets and balustrades have been added to many buildings, most notably those in a Classical style.

FIG 5.1: The parapet and entablature around the eaves of the roof were the focus of decoration on even the most restrained Classical buildings. The Victorians however went to town with this and few are as beautifully detailed as the cornice, frieze and architrave around the top of the 1862 Grosvenor Hotel, Victoria, London.

Castles and many buildings of the medieval period are distinguished by the crenellated parapets we more commonly call battlements which lined the top of the walls. Their toothed form which was originally designed to protect defenders remained fashionable on buildings erected in the more peaceful times of the later medieval and Tudor periods because of their association with power and strength. However as Classical styles became dominant during the 17th century they were rarely used. In the mid 18th century after Batty Langley published *Gothic Architecture, improved by Rules and Proportions* in which he imposed symmetry and proportions upon medieval forms to create whimsical buildings and fake ruins, crenellated parapets once again

FIG 5.2: Battlements *comprise of crenels (the gaps) and merlons (the raised section) which form the crenellations. Original examples were relatively plain in form with the occasional arrow slit as at Conwy Castle pictured here in the background. Later medieval types often had a deep overhang with corbels (brackets) and machicolations (gaps) through which projectiles could be dropped on attackers. This feature was applied to many 19th century buildings as on the gateway to Stephenson's Conwy Bridge in the foreground.*

FIG 5.3: Many *medieval churches had battlements or parapets fitted when a new lower pitched roof was built. Many were plain but others had elaborate decorative patterns as in these examples.*

45

 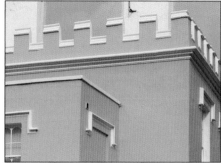

FIG 5.4: Battlements *were a feature of late 18th and early 19th century Gothic buildings inspired by Horace Walpole's Strawberry Hill (left).*

appeared on the follies erected in country house gardens. The Gothic made fashionable at Strawberry Hill inspired some to add crenellations along the eaves and angled up the gables of their picturesque houses and cottages in the Regency period. During the Napoleonic Wars patriotic gentlemen built themselves new country houses in the form of castles with battlements,

FIG 5.5: The Great Fire of London *in 1666 finally forced the authorities to take action to reduce the spread of fire between properties. The regulations that were laid down in the wake of this tragedy and over the following century forced builders in the capital to use brick and stone, fit parapets above the top floor and ensure that party walls between properties could stop flames spreading. The most visible sign of this latter action was the extension of the dividing wall above the roof line. This feature, which is characteristic of Georgian and Victorian buildings in London, was also adopted later in other towns and cities as those pictured here in Bristol.*

adding a fortified roof line which could be found on similar properties throughout the 19th century.

The parapets around Classical styled buildings were part of the carefully designed and proportioned facade of a building. In the late 16th and early 17th centuries parapets were built around the top of some of the finest houses with sections of open fretwork in geometric patterns known as strapwork, a style which was also revived in the 1830s and 1840s. Baroque architects used Classical forms and motifs for dramatic effect and many of the finest late 17th and early 18th century buildings have balustrades with urns or statues mounted along the top. More modest buildings usually had a plain or panelled parapet capped with a line of coping stones. The Palladian style which was influential during the Georgian period was far more

FIG 5.6: Some grand *Elizabethan and Jacobean houses were influenced by the latest Renaissance styles from Northern Europe. A distinctive feature of these were parapets with open fretwork in geometric shapes known as strapwork which was fashionable between around 1580-1620. This style was revived in the early Victorian period by architects like Charles Barry who designed Highclere Castle, pictured above, with strapwork around the top of this house, which was used as the location for Downton Abbey.*

FIG 5.7: It was *common in the 18th century for old stone and timber houses with steep pitched roofs to be refaced with a Classical style facade and parapet hiding the old fashioned roof from view. You can usually recognise where this has happened by looking down the side of the building as with the above examples.*

FIG 5.8: Baroque style *parapets tended to have balusters, some capped with urns or statues (top three). Modest Georgian buildings tend to have parapets which are plain, either with or without panels (bottom two).*

restrained and parapets were generally plain or were omitted altogether. In the second half of the 18th century new studies of the Ancient World uncovered a wider range of sources and inspired a variety of new adaptations of Classical architecture, some without parapets others with balustrades and statues. Throughout the 18th and early 19th centuries most urban buildings, especially those in areas which had to conform to fire regulations, had parapets along the top of the facade and in many towns and cities above the roof line along the dividing party walls.

The point where the parapet in effect meets the top of the external wall is marked on many Classical buildings by a horizontal cornice, frieze and architrave which together form the entablature. These were designed using the mathematical proportions of the Classical order from which they were derived based upon dimensions recorded on Roman and Greek temples. Whether the Doric, Tuscan, Ionic, Corinthian or Composite were used could determine the dimensions and style of the entablature. On the more restrained buildings designed in a Palladian or Neo Classical style the entablature was one of the few points where decoration was acceptable. With more modest buildings the cornice alone might be used. Early ones were made from timber but this was banned in London by the 1707 Building Act and later elsewhere so stone or other non-flammable materials were used or the cornice was omitted altogether.

FIG 5.9: Early Victorian *buildings still had parapets, some with strapwork and Gothic type designs (top row). Parapets did not become common on large buildings again until the Edwardians revived the Baroque style (bottom row).*

FIG 5.10: There were *a wide variety of parapets used in the inter-war years. Stylised Classical (top left) or Egyptian (top right) forms were popular in the 1920s whereas the simple stepped skyline (bottom row) was distinctive of the period as a whole.*

FIG 5.11: The cornices on most Classical buildings followed the forms observed in the Ancient orders although the Victorians were happy to adapt these (top right and centre). Carved scroll brackets were popular in the late 17th and early 18th centuries (top left) and along with castle machicolations (centre left) were revived on many Victorian buildings. Regency cornices were generally simple in design (top centre) but the Victorians' could be a riot of colour and decoration (centre right and bottom centre). Arts and Crafts buildings revived the use of coving (bottom right).

THE END OF THE ROOF

Gables and Bargeboards

With a simple pitched roof the two sloping sides finish on the triangular end wall, called a gable. Usually the roof extended over the gable wall but in some locations, usually where the covering was vulnerable to wind damage, the gable was extended up by a foot or so to protect the ends. In medieval, Tudor and later Victorian buildings, especially those on narrow urban plots, the gable was a prominent feature spanning the front in a single large or series of triangles. In other cases it was just a cap above a bay window. Although the

Classically inspired buildings from the late 17th to early 19th centuries had their roofs hidden behind a parapet there were cases where a shallow pitched gabled roof was used and ended in a pediment. Throughout history this section of the structure has been the focus of decoration. Timber framed patterns with hand carved details, colourful brickwork patterns and terracotta mouldings have enhanced this feature. Even some of the more restrained Classical buildings could have pediments enriched with carved stone sculpture or coats of arms.

FIG 6.1: Gable ends
not only gave builders the opportunity to create decorative patterns but also added variety in form to the facades as pictured here at Burford, Oxon.

51

Gables

In tightly packed medieval towns and cities often confined within defensive walls the individual plots of land were typically laid out in narrow strips known as burgage plots, with one end facing the main street or market place. The roofs which covered the long narrow buildings on these plots ran down the length of the structure usually leaving a gable end facing the street. With the ground floor taken up by a shop and the first floor featuring windows to illuminate the main rooms inside, the gable was the best place for the builder to display his art and timber framed decorative work is often found here especially on 16th century examples when wealth was indicated by

FIG 6.2: The structure *of a timber framed building could have additional pieces added within the gable for decoration. In the North and West of the country decorative panels were common (top) while close studded verticals were a popular display of wealth in the south and east (bottom).*

FIG 6.3: Shaped gables *became a popular feature on buildings in the late 16th (top left) and early 17th century (top right) especially in the east and south of the country. The Dutch gables with alternate triangular and segmental arched pediments (bottom) have oval plaques below which are characteristic of the late 17th century.*

the richness of pattern or how tightly packed the oak timbers were.

The influence of architecture from the Low Countries resulted in shaped gables with a series of geometric curves and angles enhancing the front of new brick buildings in the first half of the 17th century. Some of these were crowned by small pediments more correctly termed Dutch gables. Another form which was common in some regions throughout this period and into the following

FIG 6.4: Many large houses *from the late 16th through to the early 18th century had a series of two, three or more gables along the front. It was also common at this time for the width of the windows to increase from the smallest in the gable down to the widest on the ground floor.*

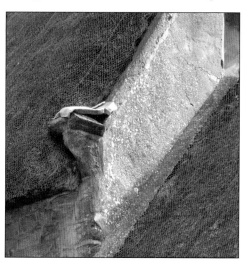

FIG 6.6: The gable end *wall was sometimes extended up above the roof to protect the covering from wind damage (top). Crow steps (bottom) can be found on some 17th-century buildings, particularly in Scotland and the east of England.*

FIG 6.5: In the east *of the country the brickwork of the gable could be set at an angle so the ends of the bricks faced up, called tumble in brickwork.*

FIG 6.7: Pediments *were shallow pitched gables which featured on Classical style buildings. Earlier types from the 17th century often featured sculpture or coats of arms within them, a fashion which was revived in the Regency and early Victorian periods (top). In the early 19th century the simple form of the pediment was used on many types of building, sometimes broken (centre top) and often with a window (centre bottom). Deep extended verges and eaves (bottom) were fashionable on Neo Classical and Italianate buildings in this period.*

FIG 6.8: Many Regency *houses were covered in stucco (render), perhaps with the top of the wall topped by battlements or shaped gables (top). Crow or corbie steps were revived in the early 19th century (centre) along with Tudor style gables with diaper (diamond patterned) brickwork (bottom).*

FIG 6.9: With the Gothic Revival gables became a prominent part of the design of buildings. They were typically steep pitched with fancy bargeboards and finials, pointed arches and decorative brickwork or timber framed patterns.

55

century had a series of small triangular gables along the front of a large brick or stone building.

As the influence of Classical styles spread so the gable as a focal point faded. Pediments were a common feature of the finer buildings of the 18th and early 19th centuries, some fronting a short gabled roof above the entrance steps, a portico, others spanning one end of a building which imitated the temples of the Ancient World. The early types which can be found on Baroque and some early Palladian buildings often had rich sculptured forms within their triangular edges, many with a coats of arms. In the mid-18th century they tend to be plain before decorative forms inspired by the more delicate artwork discovered in Ancient Greece became fashionable.

It was with the Gothic Revival that the gable became a key architectural feature once again. Now the roof was fully exposed it could be positioned asymmetrically to create a picturesque effect and decorated with all manner of patterns, mouldings and carvings. The gable, along with carefully positioned towers and chimneys, gave these new buildings a lively and balanced profile. Gables built from the 1850s to 1870s tend to be steep pitched with rich timber decoration and carvings. During the 1880s shaped and Dutch gables were features of the fashionable Queen Anne style, often with terracotta mouldings which were used to enhance the gables of buildings into the Edwardian period. In the last few decades of the century

FIG 6.10: Black and white painted timber framing was a speciality of Cheshire which the Victorians popularised partly because tar and bitumen were now available to protect old oak frames. Black and white patterned gables above red brick walls are distinctive of many buildings from the second half of the 19th century.

FIG 6.11: During the 1880s shaped gables became fashionable on Queen Anne style buildings. Some were restrained, others were packed with terracotta, stone and cut brick patterns (top row). Strong vertical bands of brickwork are also distinctive of this decade (bottom left). Arts and Crafts inspired buildings revived vernacular building styles and materials (bottom row).

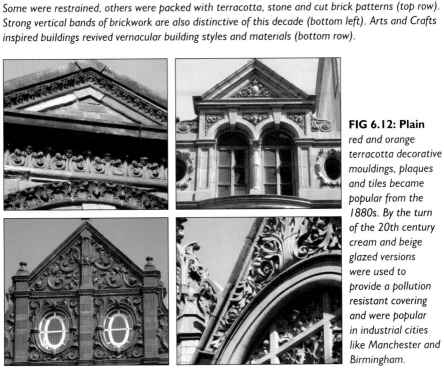

FIG 6.12: Plain red and orange terracotta decorative mouldings, plaques and tiles became popular from the 1880s. By the turn of the 20th century cream and beige glazed versions were used to provide a pollution resistant covering and were popular in industrial cities like Manchester and Birmingham.

FIG 6.13: Hanging tiles, *mock timber framing and plain bargeboards were a common feature of Edwardian buildings and on inter-war semi-detached houses.*

the influence of the Arts and Crafts Movement made timber framed, pargetting, hanging tiles and plain roughcast finishes fashionable. These forms were readily copied by speculative house builders well into the 1930s.

Bargeboards

Bargeboards are wooden planks which protect the exposed ends of the roof timbers. They have been applied to buildings ever since gables were formed and there are a few original examples on some of our oldest buildings. However most of those we see today date from the 19th century. The Victorians loved bargeboards and carved or machine cut them into decorative profiles or fancy fretwork with finials above and pendants below the top. They not only designed them to fit on their new Gothic Revival buildings but also added them to older structures to, in their eyes, improve them even though they may not have been historically correct. In general the early examples from the 1850s and 1860s tend to have

FIG 6.14: Coats of arms, *company names, the initials of builders or the date when a building was erected or rebuilt can often be found carved or moulded within the gable. They can give vital clues when trying to establish when it was built and by whom.*

FIG 6.15: Bargeboards *tend to be highly decorative with fretwork designs and deep mouldings as with the above examples. Later examples are usually plain or have simple geometric patterns.*

deeper and more elaborate patterns while those fitted towards the end of the century are plainer in form. Timber work on Victorian houses including the bargeboards and window frames would either be left to weather if it was a wood like oak, coated to appear like a fine timber or painted typically brown or dark green. Queen Anne and some later Arts and Crafts buildings used white paint for exterior timberwork which was a refreshing touch but not very practical with the sooty atmosphere, so most bargeboards would have still had a dark colour well into the 20th century.

FIG 6.16: The keeping *of pigeons or rock doves in Britain can be traced back to at least the Normans. Dovecotes still stand which date to the medieval period. They were bred for their young, called squabs, which were taken out of the nest holes when they were around four weeks old to produce a fine meat for the tables of the wealthy and as such the keeping of them was a privilege for the lord of the manor. From the early 17th century the right to keep pigeons was extended to all freeholders and pigeon holes were built into the gable ends of buildings as lofts were converted to house these profitable birds. This was not*

just to supply meat but also feathers for bedding and their droppings as a nutrient rich fertilizer and source of saltpetre which was used to make gunpowder. The keeping of pigeons declined during the Napoleonic Wars as their consumption of grain threatened supplies and dovecotes and lofts quickly became disused, although some were still kept as a quarry for shooting parties which were fashionable in the Victorian period. Other important visitors to our lofts which can gain access through tiny gaps in the gable end, verges or eaves are bats. The most common species include pipistrelle bats which tend to like tight gaps between the roof covering and underfelt, while long-eared bats prefer to be in the loft space, especially along the ridge. Most bats only use the loft for breeding during late spring and summer while they look after their young. As they are an endangered species, bats are protected by law and it is illegal to remove them or block access to their roosts.

CATCHING THE RAIN

Gargoyles and Guttering

The form of a roof and the material it was covered in were chiefly determined by the need for it to keep rain and melting snow away from the building below. However once it had performed this task what happened to the water? Today guttering sends this down underground drains but our ancestors were not always so fortunate.

Eaves

Medieval roofs were simply designed to remove the water and ensure it did not pour down the front walls to create damp inside, wash out mortar or damage the footings and foundations. Hence the roof would have projected out so that rain and snow were thrown clear, with the area close around a building where it would drip being

FIG 7.1: Having guttering which could take rainwater from the roof down into drains was such a novelty in the early 18th century that the lead rainwater hopper at the top and vertical down pipe were decorated and positioned prominently at the front of the building, as on this fine example in Ludlow dating from 1728.

61

known as the eavesdrop. People who used to stand in this area to listen in to conversation within the building were known as eavesdroppers which is where we get the modern verb 'to eavesdrop' from. Thatched roofs still have deep eaves rather than guttering with a tilting board added at the bottom of the rafters so the lower edge of reed or straw is directed up and out to keep the emerging rainwater away from the wall.

Gargoyles

On medieval churches which had parapets around the tower and along the nave the rain which collected behind these had to be discharged away from the walls. Holes were made in the parapet with a lead spout to direct the water away. It became fashionable for the outer section of this to be set within a hollow carved stone beast or figure known as a gargoyle, from the French word *gargouille* meaning throat or gullet.

Guttering

Guttering had been devised by the Romans who used lead to make water pipes; the Latin word for this metal, *plumbum*, is the origin of the term

FIG 7.2: A gargoyle *is a medieval stone rainwater spout used principally on churches and large stone buildings in the days before guttering. It was usually shaped in the form of a beast, animal or figure and many were either humorous or were designed to ward off evil spirits. Similar features which did not serve as rainwater spouts are correctly termed grotesques. The Victorians restored and created new gargoyles on their gothic buildings, these tending to be sharper in detail, less stylised and more realistic in form.*

plumbing. The Normans also used it on the White Tower, the keep at the centre of the Tower of London, which has the first recorded use of guttering when shortly after the Tower received a protective coat of whitewash in 1240 the keeper was ordered to install down pipes to prevent damage to the gleaming walls.

Early types of guttering which ran around the eaves of building could be carved out of stone as part of a cornice or parapet, or made from timber with these crude wooden troughs lined with lead where possible. This material was expensive and limited in supply until the Dissolution of the Monasteries in the 1530s when the subsequent stripping of the material from the roofs of old abbeys put larger quantities onto the market. Lead could easily be formed into any part of the guttering and was used on the finer buildings of the 17th and 18th centuries for rainwater

FIG 7.3: The rainwater hopper collected the water flowing out of the hole in the parapet so that the down pipe would not become overwhelmed and overflow in heavy rain. They were something to show off about on early Georgian houses so were prominently positioned at the front of the house as in the above examples. The best were made from lead and featured intricate designs often incorporating the date and initials of owner.

FIG 7.5: Victorian Gothic *style rainwater hoppers, straps and down pipes. These were mass produced in cast iron with gothic motifs, castellated crowns and spiral patterned pipes popular forms.*

FIG 7.4: Even the *straps which held the down pipes onto the wall could be decorated or have the date cast upon them. Drains were crude and limited at the time so the gutters may have only directed the water out into the streets. Large scale public rainwater drain and sewerage systems were not developed until the mid-19th century.*

hoppers and down pipes. As having this modern form of guttering on your property was something to show off about these features were placed in a prominent position usually down one or both sides of the facade. They were also decorated typically with the date when the building was built (or modernised) and sometimes with coats of arms, initials of the builder or owner and a range of animals and decorative patterns. Even the broad straps which held the pipes onto the wall could have motifs stamped into them.

The increased production of cast iron from the late 18th century revolutionised guttering and made mass produced hoppers, pipe lengths and brackets widely available and much cheaper than their lead counterparts. These could be cast with decorative details but by this time drainage was becoming common and less attention was paid to this utilitarian feature, so often builders would try and hide them from view. One of the advantages of the butterfly roof which was widely used on terraced housing during the late 18th and early 19th centuries was that the rain could be collected and sent to a gutter at the rear of the property so there were no pipes on the front.

One of the edicts of the Gothic Revival was honesty in building and the fact that the functional elements should not be hidden or disguised but exposed and decorated. Hence in the second half of the 19th century cast-iron rainwater hoppers with tiny battlements, gothic motifs and other decorative patterns cast upon them became a feature of fashionable buildings, with some later Arts and Crafts architects reviving the use of lead to create intricate designs.

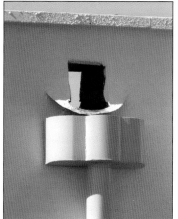

FIG 7.6: The Arts and Crafts Movement *influenced the design of some later Victorian hoppers with intricate handmade designs (top row and middle row left). In the early 20th century wavy Art Nouveau types can be found (middle row centre and right) and in the 1930s Art Deco designs (bottom right).*

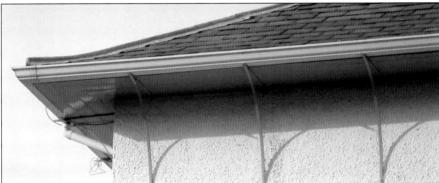

FIG 7.7: On many *Arts and Crafts style buildings the gutter along the eaves was held out on delicate and elongated arched brackets (top and centre). These deep eaves could also be supported on wooden brackets (bottom). Note that the gutter in these examples is not the usual half moon profile but is angular in form, a style which is distinctive of many buildings in the first half of the 20th century.*

FIG 7.8: The rainwater hopper *is not the only place around the top of a building where a date can be found. It is common to find a plaque set in the gable end, parapet and other places as in these examples often with the initials of the owner or builder. The date upon it may be an accurate record of when the building was erected, however in the 18th and early 19th centuries it was common for a building which was old to have the date of a modernising refit added as it was more fashionable for a commercial property to be seen as new and up to date rather than rustic and old. The Victorians became obsessed with displaying the date of construction on their buildings and most of the examples you will see today are from the 1850s through to the First World War. It is interesting to note the style of the numerals and text reflects changing fashions, so by matching the given year to a type of font you can date carved text on other buildings.*

67

THE GRAND CHIMNEY
Chimneys, Stacks and Pots

When looking up at buildings the most prominent features which are virtually universal to all old structures are chimneys. These vertical stacks of brick or stone which draw the smoke from open fires and ranges can be found in a wide variety of shapes, sizes and styles and were fitted to most houses from the late 17th century up until the 1960s.

Chimney stack

The chimney stack works because hot air rises and as it does so it draws in cold, denser air at the base. The greater the difference in temperature between the air leaving the hearth and that outside the stronger the draught is. The effect is also enhanced by making the flue longer so a taller chimney stack should perform better than a shorter one. The performance of the fire and chimney will also be improved if the flue is well insulated to maintain the heat differential. Chimney stacks that go up the outside of a building will cool the hot gases inside slowing their progress. Builders in the past did not understand the science of how they worked and although through trial and error satisfactory designs were made some of the finest period houses can still have issues with the draw of their open fires. It was not until the second

FIG 8.1: The Victorians *grabbed the humble chimney and turned it into a key architectural feature. They reinterpreted old styles, added colour, improved the performance and topped them off with all manner of interesting designs of chimney pot as in these prominent examples from Nottingham.*

half of the 18th century that men like the American scientist and politician Benjamin Franklin, who was referred to as the Smoke Doctor, and physicist and inventor Sir Benjamin Thompson, better known as Count Rumford, studied the workings of open fires and recommended changes to improve their performance. Although many of these were slow to be adopted in Britain, by the mid-Victorian period a better understanding of the importance of the size of the opening and flue, the introduction of register grates to control the air flow and the widespread adoption of chimney pots to enhance the draw made them far more efficient. From the 1930s the introduction of gas and electric heating reduced the demand for open fires and despite chimneys still being built on houses after the Second World War they often had just a single flue serving a fire in the main room.

Chimneys

The first chimneys which still survive generally date from the 14th and 15th centuries although fragmented examples from earlier ones can be found in old castles and abbeys. It became fashionable with the wealthy to divide up their old open halls horizontally by inserting a floor and creating a great chamber above, so the old central hearth had to be repositioned onto the wall with an individual chimney constructed above for each fireplace. Larger Tudor buildings are characterised by rows of tall chimneys in stone and brick, the finest examples of which would have a shaft with spiral, zig zag and diamond patterns and a decorated crown. In the 17th century fireplaces and chimneys were adopted further down the social

FIG 8.2: The Victorians' *love of decorating even the most functional of structures is best seen at the top of industrial chimneys. Colourful brickwork patterns, fancy ironwork and balconies can all be found as in these examples. The crown just below the top of the flue was not just a decorative band but a protrusion designed to stop smoke pouring down the front of the stack.*

FIG 8.3: Many early *chimney stacks were built as a separate structure on the outside of new buildings or were added to existing ones as with these examples at Little Moreton Hall, Cheshire.*

FIG 8.4: On Tudor and Elizabethan *buildings each flue usually had a separate chimney which could be set diagonally as in these examples or could be polygonal or cylindrical with decorative brickwork.*

ladder and many new houses had a back to back hearth in the centre of the building with distinctive stout, square planned chimneys emerging from the centre of the ridge.

By the beginning of the Georgian period chimneys were common even on the cottages of the poor so there was little effort made to decorate them. In fact, as Classical styles dominated

FIG 8.5: In the 17th century *it was common in many areas for a stocky chimney stack to be built in the centre of farmhouses and similar large buildings serving back to back fireplaces below.*

FIG 8.6: Chimneys *tended to become plain rectangular structures by the end of the 17th century although on Baroque buildings they could still feature some decorative effect (top left). During the 18th century they were generally plain (top right) with at best a chamfered crown (bottom).*

FIG 8.7: From the Regency *period through to the end of the 19th century Tudor and Elizabethan style chimneys were revived, especially in the 1830s and 1840s. These new versions were based upon the finest decorative originals with spiral shafts and elaborate crowns.*

architectural fashion it was preferable to hide the chimney from view so most were a plain rectangular form with a simple square or chamfered crown. The revival of Tudor and Elizabethan styles in the 1830s and 1840s resulted in the creation of wonderfully decorative chimney stacks mimicking those from the 16th century.

As the Gothic Revival began influencing architects from the 1850s so tall and prominent chimneys once again began to adorn buildings and became an important part of the design. The English architect Augustus Pugin liked the idea of having the stack built on the outside of the wall as it had been in many early buildings although this was found to cause problems with the draw as an exposed chimney stack cools faster than one built within the wall. Most Victorian architects based their designs on the chimneys from 16th and 17th century houses. Those built in the 1860s and 1870s often have decorative bands of different colour brick. In the

1880s and 1890s it became fashionable to have vertical bands of protruding brick forming patterns down the chimney with the brick all of the same colour, usually rich or dark red. Arts and Crafts architects used simple

FIG 8.8: In the 1840s *plain stacks built on the outside of the building, based upon Tudor examples, can be found on many religious buildings (left) while in the 1850s the Italianate style was very popular (right).*

FIG 8.9: During *the 1860s and 1870s colourful brick and stonework designs were popular (top row) while many Gothic Revival buildings had plain brick chimneys in a wide range of forms (bottom row).*

vernacular forms on their buildings and many of these had tall stacks based on Tudor examples or had the roughcast rendering of the exterior extended up around the chimney, which could be circular in form based upon traditional types from the Lake District.

Edwardian chimneys reflect the wide range of styles which were prevalent in the period, although most tend to be plainer in form than Victorian types. It was also common for the stack to be built vertically up emerging out of the roof halfway down the slope of the roof rather than along the ridge. In the inter-war years as cleaner electric and gas heating was introduced many modern houses tried to hide the chimney so only a short and plain stack was used where possible. On more traditionally inspired buildings the chimney was simple in design, sometimes with a ring of vertical bricks around the crown.

Chimney pots

Chimney pots, tuns or cans as they are known in different parts of Britain, have been fitted since medieval times

    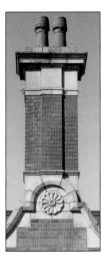

FIG 8.10: In the 1880s *prominent red brick stacks were all the rage with bold vertical lines formed out of protruding bricks. By the turn of the 20th century, designs became plain once again often incorporating terracotta (right).*

FIG 8.11: Arts and Crafts-inspired *buildings from the 1880s through into the early 20th century used vernacular styles for their chimneys as with the circular types which were a traditional form in the Lake District. Sunflowers and other foliage often featured on tiles.*

but it was in the 19th century that mass production made them a common sight. They were also added to old chimneys to improve their performance so a Victorian pot does not mean the rest of the structure is from this period. They were generally made from fired clay or terracotta, both unglazed and glazed, the latter being more resistant to pollution and water. Their main advantage is that they extend the length of the flue, and the greater the height the better the draw. This was useful when later buildings had been erected or trees had grown around a property disrupting the wind flow over the existing chimney, and pots as tall as a person have been fitted to resolve this problem. They can also taper towards the top which,

by narrowing the opening, potentially reduces the entry of rain, restricts down draughts and accelerates the rising warm air thus improving the draw. There have been hundreds of different designs over the centuries, some made as a single piece, others with component parts which were fitted together. All manner of inventive creations were patented and produced in an effort to improve performance, a common addition for instance was to have a few rings of louvred openings around the top to create a cross draw and increase the flow of smoke up the flue. Today where old pots have survived cowls, hoods, terminals and other forms of caps are fitted to keep out nesting birds, reduce rain penetration and to vent a disused

FIG 8.12: Edwardian chimneys *were inspired by many of the same sources as were the Victorians but their form tends to be more stylised and devoid of excessive decoration. The quality of bricks reached its zenith in this period and many have survived in good condition despite being exposed to rain and pollution. It was also common for the chimney to be set halfway down the slope of the roof so the stack below was straight.*

FIG 8.13: Chimneys *in the inter-war years were generally plain. In the 1920s, stacks often had a line of vertical bricks around the top (left) while on 1930s modern style buildings they were kept low and out of sight (right).*

flue.

FIG 8.14: Examples *of the wide variety of Victorian chimney pots in clay, unglazed and glazed terracotta.*

75

FIG 8.15: Water vapour *given off by the burning fuel condenses on the cooler upper parts of the flue and combines with other by-products of the combustion process to form sulphuric acid and other corrosive acids. These cause the mortar in the chimney to decay over time. As the windward or northern side is usually cooler it is this face which is prone to attack and the chimney can therefore lean into the prevailing wind direction.*

FIG 8.16: It is important *that the top of the chimney should terminate above the highest part of the roof so that it is not affected by turbulent wind eddies. The left-hand example here will struggle to work efficiently whereas the set on the right have been built with extra height probably due to tall buildings nearby. The straighter the route the better the draw and many Edwardian properties had their flues running vertically and emerging halfway down the slope of the roof rather than having the more complex tuning-fork shaped route used in earlier buildings. Old buildings usually had vents or gaps to allow cold air from outside to be drawn into the fire which is why you get a cold draught on the ground floor when a fire is lit as the chimney is drawing in cold air from outside.*

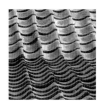

GLOSSARY

Architrave: The lowest part of the entablature and the moulded surround of a doorway or window.

Ashlar: Blocks of smooth stone masonry with fine joints.

Atrium: A top-lit court rising through a number of storeys.

Balustrade: A row of decorated uprights (balusters) with a rail along the top.

Bargeboard: The sloping boards under the verges which cover up the ends of a gable roof. Can be plain or highly decorated.

Battens: The thin strips of wood nailed across the rafters upon which the tiles or slates are hung or fixed.

Bay window: A projecting window rising from the ground, which is square ended or has angled sides.

Bonding: The way the headers (short end of a brick) and stretchers (long side of a brick) are arranged in a wall.

Butterfly roof: A roof with a central gulley running from the front to the rear of the building with the roof sloping up either side. In effect an upside down pitched roof.

Capital: The decorated top of a column.

Casement: A window which is hinged at the side.

Castellated: A battlemented feature.

Collar: A horizontal beam between rafters which is usually positioned a third to two-thirds of the way up to allow headroom in the loft space.

Colonnade: A row of columns supporting an entablature.

Cornice: Top section of entablature which also features around the top some exterior walls.

Console: An ornamental bracket.

Crown: The protruding ring of brickwork or stone close to the top of a chimney which helps prevent smoke pouring down the front of the stack.

Cupola: A small domed round or polygonal tower which stands on top of a roof or dome. Is often used as a general term for any similar open rooftop feature whether it has a dome or not.

Dormer window: An upright window set in the angle of the roof.

Double pile: A house which is two rooms deep.

Eaves: The roof overhang projecting over the wall.

Entablature: The horizontal feature supported by columns in an Ancient temple.

Finial: A vertical carved or turned feature which sits at the end of the ridge of a roof usually above a gable end.

Flashing: The metal strip around the base of a chimney, dormer window and junction of roofs to seal the joint and add support. It is traditionally made from lead or zinc.

Frieze: The central band of the entablature and a decorative band across the facade.

Gable: The triangular-shaped top of an end wall between the slopes of a roof.

Gabled/pitched roof: A roof with a slope on opposite sides with walls at each end (gables).

Hipped roof: A roof with a slope on all four sides.

Lantern: A round, polygonal or square vertical feature on top of the ridge with glass sides which allow light into the interior.

Louvre: A slatted opening for ventilation.

Mansard roof: A roof with a steep-sided lower section and low-pitched top part.

Moulding: A plain or decorative strip raised above the wall surface.

Orders: The different styles and proportions of the plinth, column and entablature from Classical architecture.

Parapet: A low wall running along the edge of the roof above the main wall.

Pediment: A low-pitched triangular feature on the top of a portico, doorway or wall.

Pendant: A hanging carved feature in the case of roof usually found beneath the junction of bargeboards, sometimes with a finial above it.

Pitch: The degree of slope of a roof. A steep pitched roof can be set at an angle of 40-60 degrees above the horizontal whereas a shallow pitched roof might be at 20-30 degrees.

Purlin: The longitudinal beams which run the length of a roof and help support the rafters.

Rafter: The timber beams running from the ridge down to the wall which form the slope of the roof.

Ridge: The top or crest of a sloping roof. The ridge piece is the longitudinal timber which forms the ridge.

Sash window: A window with two framed sashes, one or both of which slide vertically.

Segmental arch: A bow-shaped arch which is formed from a segment of a larger arch.

Skylight: A window set in line with the roof covering or formed into a low glass roof to illuminate the interior.

Stucco: A smooth plaster rendering which imitated fine cut stone.

Truss: A framework of timber or metal beams which form triangles to support a structure like a roof or bridge. A triangle is the only geometric shape which cannot be distorted by external force or load.

Verge: The sloping edge of the roof above a gable end.

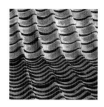

INDEX

OTHER TITLES FROM COUNTRYSIDE BOOKS

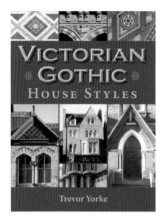

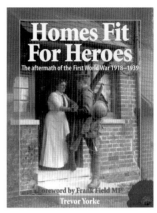

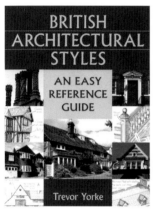

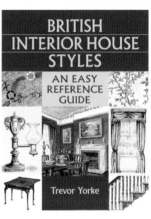

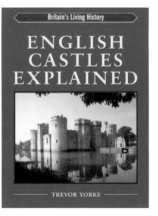

To see our full range of books please visit
www.countrysidebooks.co.uk

Follow us on @ CountrysideBooks